Artificial Body Parts

by Kathryn Stelljes

Printed in Mexico

ISBN 978-0-15-364954-7
ISBN 0-15-364954-2

2 3 4 5 6 7 8 9 10 126 16 15 14 13 12 11 10 09 08

Visit *The Learning Site!*
www.harcourtschool.com

Introduction

An engineer is someone who builds, designs, or manages machines, engines, roads, bridges, railroads, mines, electrical systems, and almost any other thing involving technology and mechanical devices. A doctor is someone who treats physical diseases and injuries. It's not often that you think of these two people working together toward the same goal. But when the machine is the human body, both professions come together.

Simple Machines

All machines are made of one or more of six simple machines: the lever, pulley, wheel and axle, inclined plane, screw, and wedge. An intricate machine that is very close to you is no exception to this. Your body is a machine. Your bones act as levers, and your joints act as fulcrums. Your teeth act as wedges. Even your knee joint can be compared to a simple machine: the pulley.

People are, of course, more than machines. However, many parts of your body function as simple machines that can be replaced with artificial parts when they are damaged by disease or injury.

Perhaps the most common simple machine in your body is the lever. A lever is a rigid bar that turns around a fixed point. That fixed point is called the fulcrum. The fulcrum is one of the three parts that all levers have. The bar of a lever has the other two parts, even though the bar itself is rigid: an effort arm and a resistance arm.

There are three classes of levers. Each class of lever depends on the locations of the lever's three parts. The diagram shows what all three classes look like. As you can see, a human arm is a good example of a third-class lever.

Three classes of levers

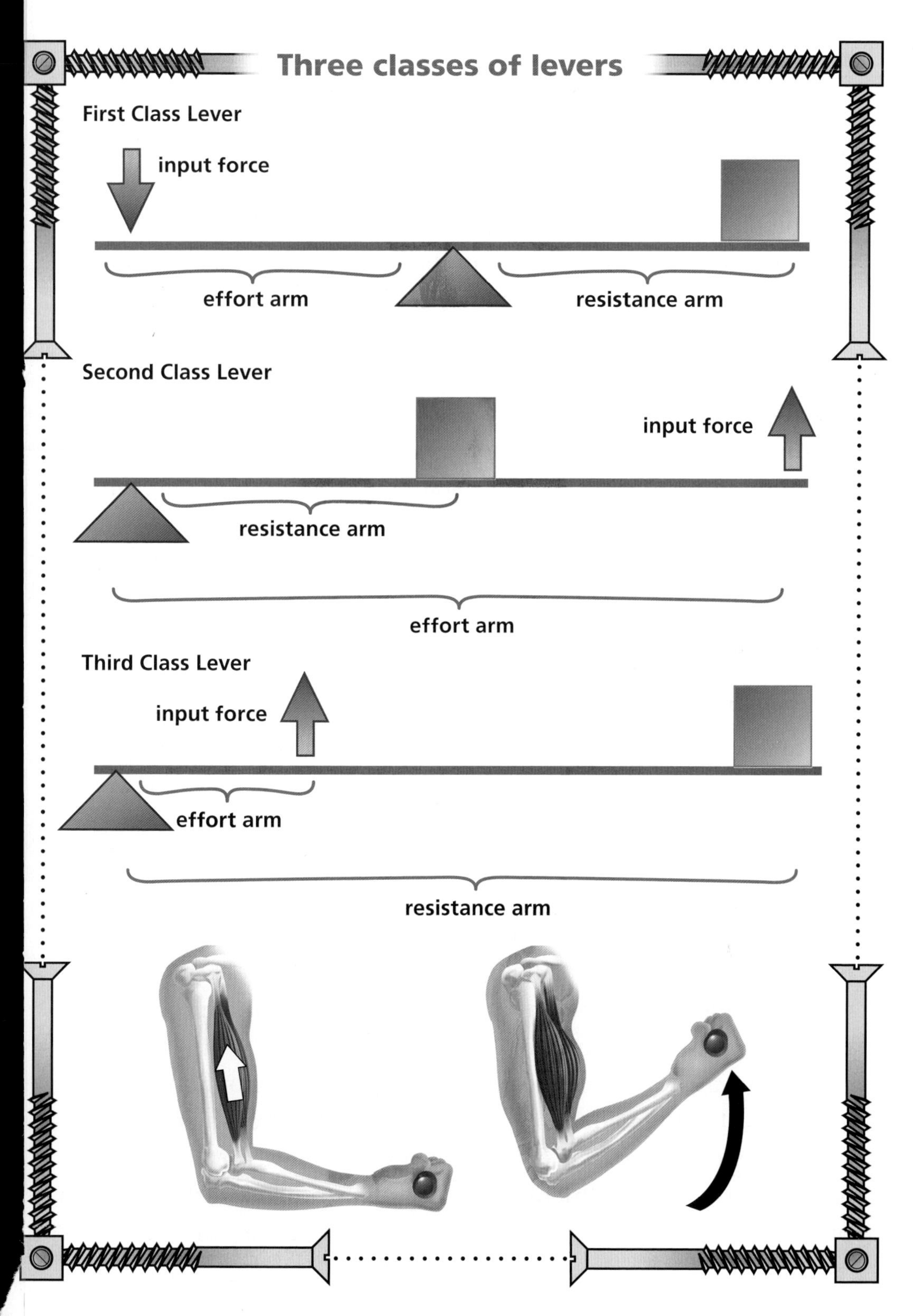

Engineering Legs

An artificial leg or arm, or any other artificial body part, is called a prosthesis, or a prosthetic device. Some of the first prosthetic devices were artificial legs.

You have probably seen such artificial legs in movies or in drawings of people in books about the 1600s or 1700s, for example, a pirate with a peg leg. Peg legs may seem to be something that the writer or artist just made up. Not long ago, however, such prosthetic devices were the best option that you had if you were unfortunate enough to lose a leg.

An artificial leg has to be strong. Your legs and feet bear the entire weight of your body when you are standing. Therefore, strong, durable materials such as wood and leather were used to fashion some of the first "peg legs." They were nothing fancy. They had only one purpose: to bear weight and make it possible for a person to get around.

Your leg and foot do more than just bear your weight, however. Your knee and ankle joints help absorb the shock of your foot hitting the ground as you walk or run. Those same joints allow your leg and foot to flex and bend as you walk over uneven ground or climb stairs. To help you walk, each of your feet has 26 bones and a flexible arch.

The best prosthetic leg and foot would, of course, mimic the function of the real thing. More functional prostheses became available in the early 1800s, when a British craftsman invented a leg with hinged metal knee and ankle joints. He attached artificial tendons that lifted the front of the foot as the knee bent during walking. That was the beginning of prosthetic legs that provided a more natural way of moving.

Today's artificial feet provide for a wide range of options. Real-looking feet with molded toes and rubber heels for cushioning give some people an attractive option. However, some artificial legs and feet are designed for athletes. These complex prosthetic devices aren't pretty; but, for a runner, they provide the best performance.

Someday scientists hope to make much more realistic prosthetic legs by using computer technology. A microprocessor implanted in the leg would guide motors and hydraulic joints in the leg to allow the person wearing the device to walk with a more natural movement, using information about the ground surface gathered by sensors in the leg.

Prosthetic legs aren't peg legs anymore.

Engineering Levers: Arms, Hands, and Fingers

Making a useful artificial arm and hand is a bit more challenging than making legs and feet. Though your arms and hands do not support your body's weight, normal hand movements, such as those needed for grasping and handling small items, are more complex and demanding than leg and foot movements.

Like the leg, the arm consists of a large upper bone connected to two lower bones through a joint. The elbow joint serves as a fulcrum that allows the arm to bend and act like a lever. The hand, too, is a lever—the wrist is the fulcrum. But the hand is also a much more intricate tool made up of several smaller levers that work effortlessly together: your fingers.

In the past, people who had lost the use of their arms or hands usually had to choose between a hook and a wooden or iron hand. A hook can be used for grasping and manipulating objects, but it does not make an attractive hand. The less useful wooden or iron hand was usually just "for show."

In time, people saw a need for something more useful. Utensils were invented for use in place of a hook. People could unscrew the hook and replace it with a fork, for example. Eventually, more advanced artificial arm and hand units were developed that had movable joints attached to the straps of a harness worn over the shoulder. By moving the shoulder or upper arm in a certain way, a person could move the artificial hand.

Thanks to engineering, prosthetic arms and hands have continued to improve. Today we have the myoelectric limb. When a muscle flexes, it gives off a small electrical signal. This tiny signal, which is caused by chemical interactions in the body, is in the range of 5 to 20 microvolts. To get an idea of how small that is, think of this: a light bulb in your house uses around 110 volts. That is more than a million times the strength of the myoelectric signal that your muscles produce. Although a muscle's signal is very small, it can sometimes be used to control a prosthetic limb.

The myoelectric limb uses an electrode to pick up the muscle's signal. Then a controller amplifies and processes the tiny signal, making it much more powerful. The amplified signal is then used to control small motors inside the prosthetic that can move the hand, elbow, or arm. The effect is that you are controlling the artificial arm and hand in nearly the same way that you would control a living arm and hand.

The Wedge

Levers aren't the only simple tools that can be found in the human body. There are also wedges, but you are more used to calling them teeth.

A wedge is two inclined planes placed back to back. The purpose of a wedge is to push things apart. When you push down on a wedge, as shown in the diagram below, the resulting forces go outward.

A knife is a type of wedge, and so is the incisor tooth. Incisors are your front teeth—the ones you use to bite into things. That's how you can make smaller pieces out of a large piece of food that would otherwise be too big to fit into your mouth.

But what if you lose your teeth? This was a common occurrence many years ago before modern dentistry and fluoride toothpaste. Until the 1890s, people were unaware that microorganisms in the mouth caused cavities. As a result, people did not know how to take

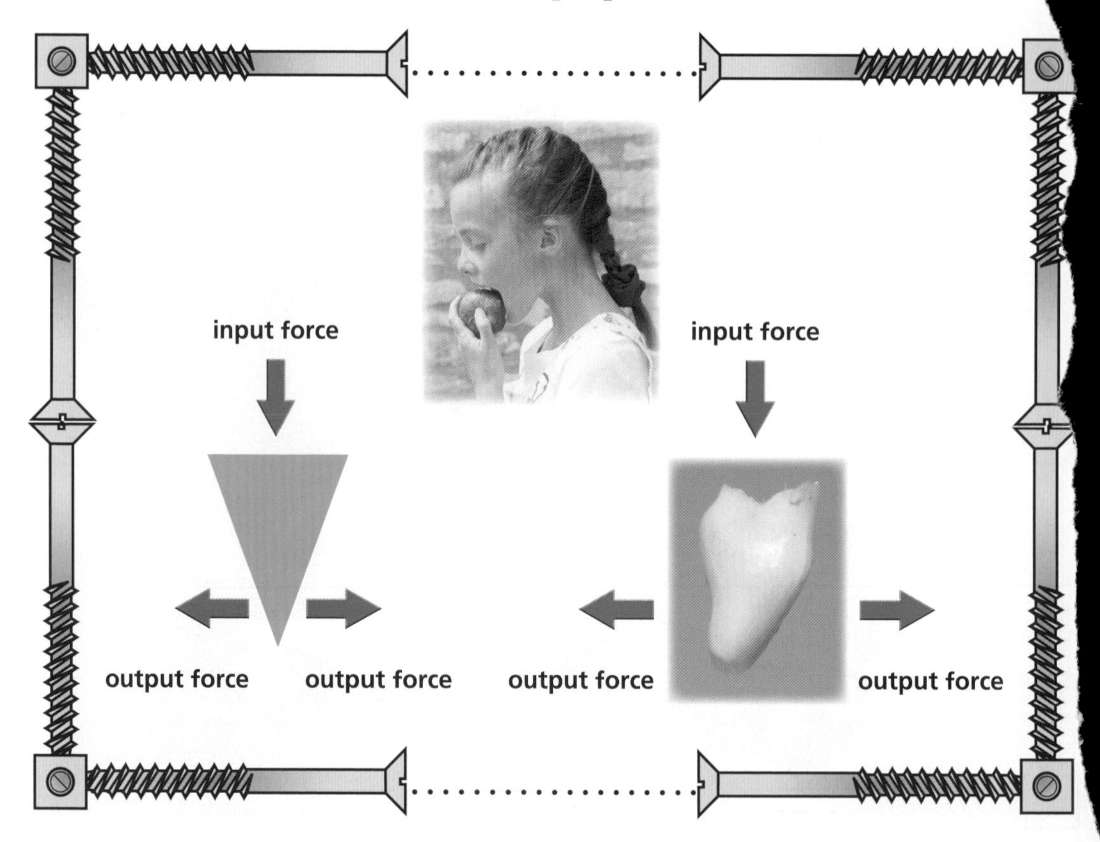

care of their teeth, and cavities were common. Tooth decay led to toothaches that could be stopped only by pulling the teeth.

Losing your teeth puts you at a great disadvantage when it comes to eating. Certainly you could cut your food up into small pieces, or you could just eat soft foods. That would severely limit the types of food you could enjoy. But there's another problem with losing your teeth—your appearance. So, even in the days before modern dentistry, people came up with the idea of false teeth.

Dentures, or false teeth, are one of the oldest and most common prosthetics. Records of teeth made of gold go back about 2500 years. False teeth were made from animal bones or ivory from hippopotamus or elephant tusks. Sometimes people even removed and used teeth from people who had died. The replacement teeth had to be tied to existing teeth with silk thread. If a group of teeth was missing, denture bases were carved or hammered to an approximate fit.

The first American president, George Washington, was one of many people in the 1700s with dental problems. He started losing his teeth when he was in his twenties. Eventually he lost all of them.

Washington's dentures

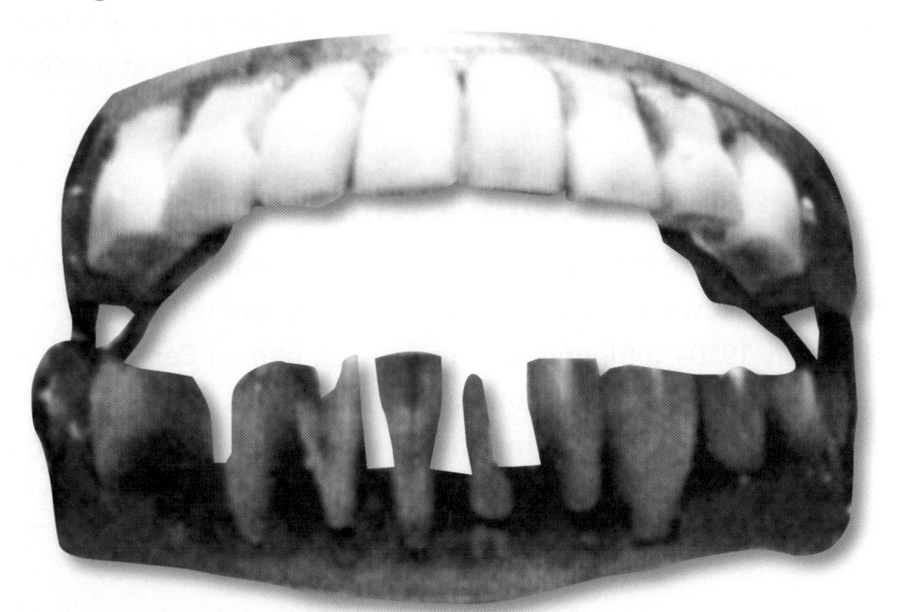

Washington met a dentist in the army who made him a good set of false teeth. John Greenwood, who was also a soldier in the Revolutionary War, fashioned a set of teeth for Washington. The base was made from a carved hippopotamus tusk. He made Washington's top teeth from ivory. The lower set was made of eight human teeth fastened into the base with gold. Springs between the top and bottom sets of teeth kept them set properly in the mouth, but they never fit or worked very well.

In 1789, the technology of teeth improved with the development of teeth from a hard, white clay mixture known as porcelain. By 1825, porcelain teeth were manufactured commercially and in large numbers, making them more affordable. But there was still no suitable base to which teeth could be attached.

Then in 1839 American inventor Charles Goodyear discovered a process for hardening rubber called vulcanization. The hard rubber could be molded to make many things—including the tires that still bear Goodyear's name. The rubber also turned out to mold well to the shape of the mouth and soon became the preferred material for denture bases.

Despite many improvements in cleaning and repairing teeth, many people still lose their teeth to injury, disease, and old age. Millions of people wear either bridges, which replace a row of a few teeth, or complete dentures. The bases are custom fitted and made of plastic or ceramics and metal.

Something Better Comes Along

A discovery in the 1950s brought another alternative for replacing teeth: implants. However, it began not with teeth, but with bone.

In the early 1950s, a bioengineer in Sweden named Per-Ingvar Branemark wanted to find out more about how bones heal. He implanted small titanium rods into living bone. The rods contained a tiny monitoring device so that he could observe the healing process

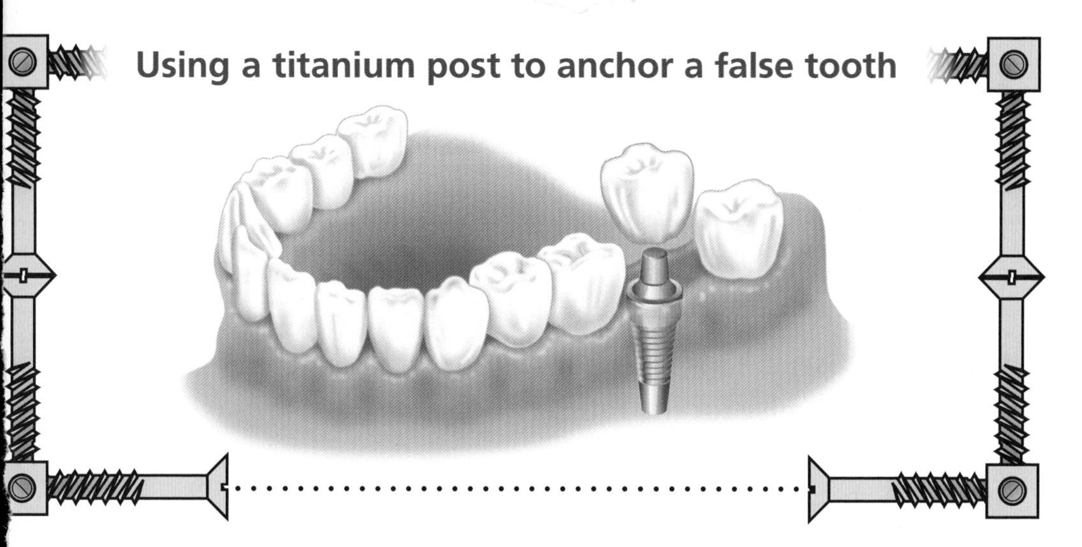

Using a titanium post to anchor a false tooth

inside the bone. When the experiment was over, however, he found that he could not remove the titanium rods without destroying the bone. The two materials—bone tissue and titanium—had become integrated, or like one material. Instead of being rejected by the body as a foreign object, a titanium implant actually encouraged the growth of bone tissue!

It didn't take long for Dr. Branemark to realize the importance of this discovery. Titanium could be used to connect a prosthesis directly to human bone. The first use of this discovery was in replacing teeth.

In the 1960s, dentists in Sweden began to experiment with dental implants that used titanium to anchor false teeth directly to the jawbone. Up to that time, someone who had no teeth had only two options: be toothless or wear dentures. Although twentieth-century dentures were greatly improved over those in George Washington's time, they still had problems. Sometimes they didn't fit properly. They would come loose and had to be glued into place. Also, because they covered so much of the inside of the mouth with foreign material, they interfered with the sense of taste. Implants offered an exciting alternative.

During the 1980s, Dr. Branemark introduced his techniques to American dentists. In time, the implant procedure was refined to be used for more than just anchoring a full set of dentures. Today, individual teeth can be replaced with artificial implants.

If it works for wedges, what about levers?

After so many years of success in dental implants, titanium is now being used to anchor prosthetic legs and arms directly to bone. This gives you more control over moving the prosthetic limb and actually allows you to "feel" with it. The sensation of feeling results from movements and compressions being transmitted directly to living bone tissue. This use of implants is still new in medicine but is showing promise for the future.

Bioengineering

Bioengineering combines medicine, biology, and engineering. Many universities now offer degree programs in bioengineering. A few of the special areas offered for study are biomedical imaging, which focuses on such tools as magnetic resonance imaging (MRI) and ultrasound; bioinformatics, which focuses on using computer technology to analyze molecular structures, including that of DNA; and biomechanics, which focuses on artificial limb and joint design. As technology advances, bioengineers will continue to work toward developing prosthetic limbs that respond directly to instructions from the brain. The first experiments in this direction are now underway.

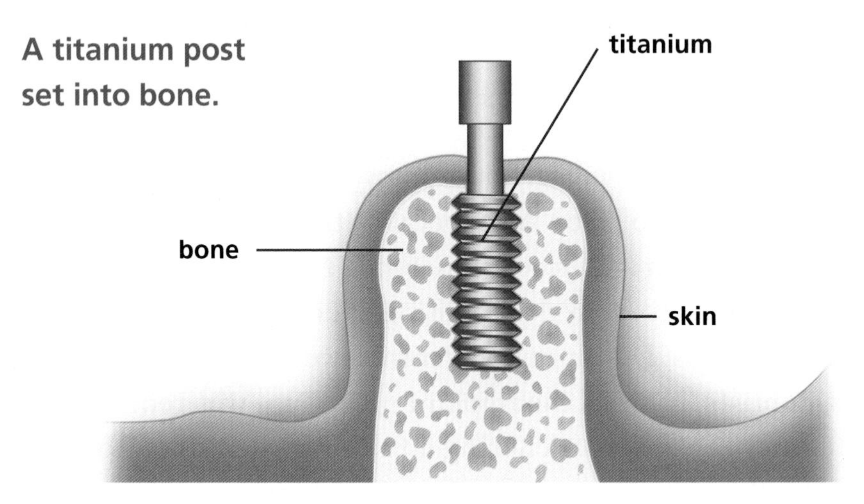

A titanium post set into bone.